Linux:

2019 Easy Guide for Beginners to Learn the Linux Operating System and Linux Command Line. 99 tips and tricks included.

Linux

Copyright © 2019

All rights reserved.

ISBN: 9781688712454

CONTENTS

Introduction 4

Chapter 1 – Linux File System 11

Chapter 2 – Linux Advantage 19

Chapter 3 – How To Set Up Linux 29

Chapter 4 – Linux Features, Components And Architecture 35

Chapter 5 – Comparison Between Linux And Other Operating system 48

Chapter 6 – Linux Tips and Tricks 55

Conclusion 74

Thank you for purchasing this book!

We always try to give more value then you expect. That's why we've updated the content and you can get it for FREE. You can get the digital version for free because you bought the print version.

The book is under the match program from Amazon. You can find how to do this using next URL: https://www.amazon.com/gp/digital/ep-landing-page

I hope it will be useful for you.

Introduction

To most people, Linux means a whole lot of things: A horrible system with complicated commands or an operating system for the experts but we like to see Linux as a basic operating system, just like Windows or macOS X. It allows you to work as you would under Windows. But it works differently.

For several years, Windows OS were heavily favoured than Linux because of their ease of use but this is no longer the case. Linux now has a nice, comfortable and easy-to-use graphical system that puts it on par with Windows operating system. The commands are more simplified and a lot of upgrades have gone into latest versions (distribution) to improve users experience.

Even though most people see Linux as an independent operating system, in the real sense, Linux is pretty much everything. Its presence is felt everywhere and almost all devices work with Linux. From your Android phone to smart fridges and even rifle used by the military.

Often referred to as the kernel, Linux performs a whole of function in the operating system. Its function involves taking care of dirty work which includes: memory management, access to peripherals (hard disk, CD-ROM reader, keyboard, mouse, graphics card ...), network management, time sharing microprocessor between programs (multi-task), etc.

Unlike Windows whose GUI is imposed on you, there are different graphical interfaces under Linux, the main ones being Gnome, KDE and XFCE. It's even possible to run Linux without a GUI, or even launch the GUI only when you want. Hence, with Linux, you have total control over the graphical user interface which is interesting for users who like to tweak things.

What is GNU

GNU is a project that has brought lots of utilities to the Linux kernel, such as the famous GCC compiler, and thousands of utilities (tar, tail, man, bash ...). These utilities give users the license to perform a whole lot of tasks which includes copying, formatting and removing files from the system by simply clicking one or two buttons without writing commands.

These GNU utilities, associated with the Linux kernel, make up the GNU / Linux operating system.

GNU / Linux is free, different companies took over and decided to forge an operating system to their liking. These are called distributions. Among the best known include RedHat, Fedora, Mandriva, Debian, Susa, Slackware, Gentoo, Xandros, Lycoris ...

In addition, this platform gives users the freedom to make a choice of software. This includes making a choice of commands as well as graphical interface. Predictably, this feature comes as a surprise to Windows users who are stuck with a default command line.

Fortunately, Linux comes with a host of advantage windows users should look to explore. First, the possibility of it crashing is less and supports multitasking. With this, you can run several programs at the same time without encountering any problem.

Linux Distribution

Linux Distributions also called distros are versions of Linux operating system that comes with Installation programs and management utilities just as we have Windows 7,8 and 10 on Windows OS.

Linux distributions which are designed with kernels are often easier for users to operate than open-source Linux which most times uses commands. This distribution eliminates the needs for users to compile existing codes before an action is performed.

As the years roll by, more distribution are developed by vendor each bringing a new feature to the table. Presently, there are thousands of distributions to choose from and they include Ubuntu, Debian, Fedora and Slackware. With millions of distributions in the market, one is bound to easily get confused on the type of distributions to choose For anyone just getting started with Linux, it's important to put price and orientation into consideration.

Orientation: For example, the RedHat are highly oriented corporate servers (databases, web servers ...), Mandriva and Ubuntu are more oriented towards office users and Internet users, Flonix is designed to start directly at from a USB key, etc.

The way they are "fabricated": for example, the RedHat is designed by a big company, while the Debian is designed more democratic (participation of Internet users).

The price: Some are paid (RedHat, Mandriva ...), others free (Fedora, Debian ...). Note that you may have to pay for free distributions, but the price is only used to cover the support (CD), shipping costs and any paper manuals. Nothing prevents you from downloading and burning them yourself.

When it comes to choosing software, it's hard to say. It all depends on your level and what you want to do with it. To know which one is better, I encourage you to download various distributions to test them and find the one you like the most.

If you do not know where to start, I recommend the following:

Knoppix

Knoppix (if you do not want to install anything on hard disk). This version of Linux starts directly from the CD and writes nothing to hard disk. No installation is necessary on hard disk. It is a way to discover Linux safely.

Ubuntu

Ubuntu is a great distribution, which can either be used as Knoppix (without installing anything) or installed on hard disk. The interface is very clean and simple to use. Once installed, you can access hundreds of additional software in a few clicks.

This is the distribution I recommend if you want to install Linux on your computer. Sometimes there are derivatives of these distributions. For example, Knoppix is a distribution derived from Debian, and Morphix is derived from Knoppix, etc. Ubuntu is the most popular Linux distribution.

Linux and other Unix

Linux was created by Linus Torvalds in response to the big commercial Unixes, which were mostly overpriced. The GNU project has also started on a similar motivation (GNU means " G NU is Not U nix ").

Linux is said to be a free operating system, which means that you are free to use, modify and re-distribute it (which is not the case with Unix, Windows or MacOS X). These Unix still exist today and are still sold: HP-UX (Hewlett-Packard Unix), AIX (IBM Unix), Solaris (Sun Unix), IRIX (Unix of Silicon Graphics) ...

Unix is a registered trademark, and any company that wants to create a " Unix " operating system must follow a number of strict rules. With its free, open and performance, Linux is gaining popularity compared to other Unix. Even the big companies that were doing their own Unix get started! (like IBM, Sun, HP, SGI ...)

Linux and Windows

I do not intend to relaunch another debate " my system is better than yours ". The rivalry between Linux and Windows System is just like the debate between iPhones and Samsung Galaxy versions. They just can't go away and users are never tired of asking for more.

These competition have seen improved versions from every manufacturer which has really shaped users experience. Here, we do not intend to criticise or crown one the winner. Because in the actual sense, both have their own use. No version is perfect. All has it's own pros and cons. The best, however, depends on the user and what you intend to use it for

On the other hand, I just want to give some elements that can help you decide between Linux and Windows.

" Decide " is also a big word, since nothing prevents you from installing the 2 on your computer and switch from one to another!

However, It must be emphasized that Linux require more time than Windows, to master. If you are not ready to spend more time, do not go to Linux.

In his defence, mastering Linux is very gratifying, because not only does it allow you to understand what is going on "inside" (if you wish), but above all to do exactly what you want. The learning curve is steeper, but it goes further.

Note that with recent distributions like Ubuntu, Mandriva or

Xandros you do not have to put your hands in the grease if you do not want it. They are as easy to use as Windows.

Chapter 1 – Linux File System

Anyone who uses a computer must inevitably use the files to do his job; whether he is a simple user of Word or Excel or that he is a programmer of software or systems. This is the most visible part of an operating system. Moreover, many people judge the capabilities of their work environment on the quality of their file system, interface, structure and reliability.

Today, interfaces that are more and more similar to each other offer us a very good ability to manipulate files, that is to say, to name them, to move them and to authorize all kinds of operations. These interfaces now hide the internal structure of the files and let us think that they are all similar, hidden behind similar folders and icons.

A file system is a data structure for storing information and organize them in files on the so-called secondary memory (hard disk, diskette, CD-ROM, key USB, etc.)

For the moment, the various operating systems adopt their own solutions for safeguarding user information; although this is almost exclusively through objects called files that may contain programs, data and other types of information.

The operating system provides special operations (system calls) to create, destroy, read, and so on.

All operating systems are striving to achieve independence from peripherals. Operating systems consider the file system essentially as an interface with relatively static objects. This interface has its own structure, often hierarchical, to designate a set of data, but also devices.

The purpose of the file system interface is to

trivialize the objects it covers as much as possible, so the Linux and Windows / Dos systems also name their objects in a uniform way.

Checking Files On Linux

Under LINUX, the command ls - is used to check the type of files: the first character of the result gives us this information: " - " means ordinary files, " d " directories, " c " character devices, " b " block

Modern operating systems adopt a hierarchical structure of files. Each file belongs to a group and each group belongs to a higher order group. These groups are called directories or folders.

The appearance of the general structure of a file system takes the appearance of a tree, formed from a "root" directory covering devices and especially one or more disks. In each of the directories, we can find other directories as well as ordinary data files.

Media types

Device names:

- FDI

/ dev / hdb8 is the 2nd IDE hard disk and the 8th

logical partition.

- SCSI and ZIP IDE

/ dev / sda1 is the first SCSI hard disk and the first partition.

- FLOPPY

/ dev / fd0 is the first diskette drive (A: \). The number of blocks on a 3 1/2 "high-density floppy disk is 1440. The floppy disk must be formatted before mounting.

- NULL

/dev/null is the garbage bin, and the files that are poured into it are irretrievably lost.

- CD-ROM

/ Dev / cdrom. There is a symbolic link between the "/ dev / cdrom" file and the "/ dev / cdrom0" device name.

The removable cartridge drives are mounted as external hard drives. You first have to partition the drive with " fdisk ", then create a partition table and install a file system with " mk2fs ", then you have to create the mount point with " mkdir " and mount the partitions with " mount " Finally, add a line in the file " / etc / fstab " so that the device is mounted at each startup.

Access rights

rights or permissions allow certain users or groups of users to restrict access to certain files or directories.

There are three types of access rights:

read (r) for reading access to the file (allows the printing, displaying and copying of a file, and allows traversing the directory or displaying files in a directory)

write (w) for write access to the file (allows the modification of a file, and allows the deletion of a file or the saving of a file in a directory)

execute (x) for the possibility of executing the file (allows the execution of a program, an executable, and allows to access the management information of the files of the directory, like the inode, the table of the rights ...).

For each file, access rights are set for three categories of users:

user (u): the owner of the file

group (g): the group that owns the file

all (a): all users

To view the rights of all files in the current directory:

ls -la

The access rights can also be expressed in their octal form, that is to say using a number from 0 to 7 (there are eight possibilities, which can be fixed with only 3 bits). Each of the rights (r, w, x) corresponds to an octal value (4, 2, 1), the octal values are accumulated for each type of user (u, g, o). For each type of user (u, g, o), the value in octal can take the values 0, 1, 2, 3, 4, 5, 6 and 7. For example, the combination of all rights cumulated for three types of users (rwx rwx rwx) is equivalent to the octal value 777.

0 means no rights

1 is the executable right (--x)

2 is the write right (-w-)

3 corresponds to the cumulative execution and writing rights (-wx)

4 corresponds to the right of reading (r--)

5 is the cumulative read and execute rights (rx)

6 corresponds to the cumulative rights of reading and writing (rw-)

7 is the cumulative read, write and execute rights (rwx)

For example:

666 gives the right to all read and write

764 gives all rights to all

700 gives all rights to the file owner

The octal number can be four digits when the super user sets the special rights ("s" and "t").

File Naming Rules Old Unixes

were limited to 14 characters, but nowadays long file names are handled from 1 to 255 characters.

The slash (/) is forbidden because it is the directory delimiter in the tree, and represents the root, ie the top of the tree. Files whose names begin with a period are hidden or hidden files, they do not appear by default with the "ls" command without the "-a" option, and most commands do not take them into account. less than mention it explicitly.

The double dot (..) identifies the parent directory and the dot (.) identifies the current directory or working directory. These two files exist in all directories. It is therefore not possible to name a file with a single point or with two points since the pointers already exist (it is not possible to have two files with the same name in the same directory, and it is not possible to you can not delete the pointer to the current directory or the parent directory).

Chapter 2 – Linux Advantage

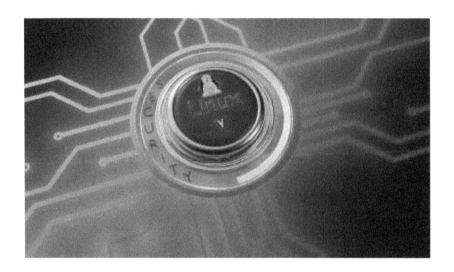

From most Computer novice, we often hear things like "Linux is complicated", or "Linux is a geek thing". If there are of course Linux distributions that only passionate experts are able to operate; There are also some that are specifically intended for the general public: Linux Mint, Ubuntu and its derivatives.

You will discover that GNU / Linux is very easy to use and much more user-friendly than what most people think.

What are the advantages of using GNU / Linux?

Hands-on your steering wheel

With this software, You are the master of the board. GNU / Linux will do absolutely nothing unless you ask it explicitly, unlike a proprietary system that decides everything - or almost - instead of the user.

You're not mandated to install what you do not want to install, nor will a completely new system be imposed on you while the one you paid is still working very well.

Conversely, you will probably not have anything to uninstall from the start; no anti-virus trial for a month, no key, no locks, in short, no bloatware.

After years of relying on Windows, the freedom that Linux grants can be scary. However, It should be understood that GNU / Linux allows you to manipulate your operating system, to make it a space arranged to your taste, your needs, as much as it gives you the right to do anything at all .

Serenity and Efficiency

Even though some distributions are complicated,

there are still some GNU / Linux distributions that are very suitable for beginners. Everything is therefore designed to make life easier for beginners.

This mean that you will not need a master's degree in computer science to operate Linux. Moreover, GNU / Linux offers a reliable environment, technically and humanly.

When you're done installing a distribution for the first time, the most striking thing about it is its barebones look. Usually, only two icons are present on the desktop: the "workstation" and a personal folder. Not even the trash! It may seem austere but it is also the symbol of the GNU / Linux experience: simplicity and the choice to add things according to your own needs.

This simplicity is immediately visible when you go around the interface. Unlike Windows, where most people who have not yet mastered the system waste time uninstalling useless software and disabling imposed functions; under GNU / Linux you don't have to do this. No need to uninstall a useless bunch of software. Rather, you choose the software to add yourself.

Precisely, nothing is easier than to install software chosen among the thousands available in a software library. No need to waste time searching the net for the desired application that could just as easily contain a virus or corrupt files in your system.

The most eloquent feature of GNU / Linux, and not least, is that no restart is necessary after installing software. This is so for updates that are operational instantly without rebooting the system. A breath of fresh air compared to the slow heaviness of Windows updates.

Security

There is no need for anti-virus when you are on Linux. This seems inconceivable but is true because this operating system is built primarily to be as safe as possible, it is part of the DNA of Unix-like systems (Linux, BSD) where security is such an important element as other components of the system.

Despite this claim, most computer operators have tried to give an explanation about the security build-up of Linux. On one hand, some operators suggest that there are viruses on Linux and therefore it is important to protect yourself. The others argue that this anti-virus only detects those intended for

Windows and that, moreover, the diversity of the Linux world means that a virus which affects one distribution may not affect the other.

The most common security requirement for a Linux system is that a password is requested as soon as you log in or when you want to perform a potentially dangerous system modification operation (update, installation of new software, etc.). The password will also be required when you want to access some files as this will prevent some people from searching your data when you're not around.

As far as updates are concerned, they seem much less frequent and in most case, much faster to perform.

We sometimes hear that Windows was designed for one thing: to work. Originally, security was not even a primary concern, an aspect so little taken into consideration that it was delegated to third-party software.

Hence the very lucrative industry of anti-virus and anti-malware. Even if one can argue that nowadays Microsoft takes serious account of security problems, the continuous and frequent flow of security updates should encourage all users to ask serious questions about the reliability of Microsoft operating system.

A **new version in a snap of fingers**

On Linux, If you want to change the version, for example, to have a most recent operating system, most distributions are to be reinstalled in full.

Under Windows, this operation causes stress and takes a lot of time. To change GNU / Linux version is so easy that it can easily become a routine.

To perform this In Linux Mint, just go to the update manager to start the installation of the latest version available.

It is also interesting to note that installing a recent version does not mean changing the whole software. GNU / Linux offers operating systems light enough to run on computers with less than 4 GB of RAM.

Generally, consider that with present Linux distributio, an entry-level computer with 4 GB of RAM is almost oversized for common uses. The minimum system recommended for installing Linux Mint is ... 9 GB of disk space and 256 MB of RAM (1GB for comfort)!

Obviously, some software will probably require a little more than 1 GB of RAM. But between 1 and 4 (or 8 GB of RAM), space is enough.

Finally, the different versions of the distributions are often proposed in "long cycles", commonly called LTS (Long Term Support). For example, Linux Mint 18 available since spring 2016 will be supported until spring 2021.

For the rest, if you know how to copy and paste, you will reinstall your software and replace with the recent versions.

Above all, it is possible to restore all the software configuration of the old system, which includes, for example, your e-mail, your settings, as well as the complete parameters of your browser with extensions included!

How is this magic possible? The advantage of a GNU / Linux distribution is that all the personal files are deposited in a single named folder. This folder also contains hidden directories and these are the ones that contain all the settings of your software.

GNU / Linux does not violate the user

GNU / Linux is an operating system that does not

exploit you. The default installation does not include much more than the most commonly used and fully functional software. The Linux community and free software developers do not care what you do with your computer for which you are solely responsible.

A GNU / Linux distribution does not hint, analyze or transmit any of your actions to "third parties", which may breach users privacy.

The **user is not locked**

Most Linux distributions are free. There are no restrictions on installation and making copies is not only legal but encouraged. That said, we can contribute financially, even modestly, and even if according to the rule of free software, nobody is obliged to the slightest ...

Therefore, say goodbye marketing jargons such as free update during "x" months or years; activation keys, licenses, questions about how to move to the higher version forcibly and as quickly as possible.

No more handcuffs! Install Linux Mint or another distribution on all the computers in your home. Install Debian on your personal computer,

HandyLinux on your children's (s), Ubuntu on your spouse's; install the previous or new version and even the future. In short, do exactly what you want.

GNU / Linux is the opportunity to work natively in a free universe, a holy environment that you control at your leisure and can modify according to your knowledge, your needs and your desires. Or do not change anything at all. In other words, you decide the direction to take and if one day you do not like the one that takes your favourite distribution, do not hesitate to uninstall them.

A **complete operating system**

The question we ask most often is whether everything we do on Windows will be possible under GNU / Linux.

The answer is yes! All distributions contain free software equivalents to those found on windows. An ultra-complete office suite, PDF reader, social media app, browsers, mail app, photo viewer, multi-media, photo editing and a whole slew of specialized software or utilities. Some even offer proprietary software like Skype or Flash for those who can not do without it.

In other words, unless you have such specific and special requirements, you will find in a GNU / Linux distribution, all the tools you need to perform all tasks, from the most complicated to the simplest.

Chapter 3 – How To Set Up Linux

By default, most of the time, when you buy a computer, an operating system is already installed. This is often Microsoft Windows. It is also possible to choose computers with Mac OS. On the other hand, the purchase of a PC with Linux as a default operating system is very rare. When you first start your brand new computer, you do not even wonder what kind of operating system it will start. Your only concern is to be sure that the PC starts and you can use it right away.

However, If your computer had the unique Linux

operating system, it would be enough to start the PC and start using it. So to summarize, the best way to install Linux easily is to make this program the only operating system available on your hard drive.

As soon as it comes to coexist with other operating systems, things start to get a little complicated. Most people will tell you that you can install Linux together with Windows using both interchangeably. We aren't arguing about that just that you will encounter one or two hitches while doing so.

Nowadays, Linux usually comes in the form of a distribution that can be placed on a DVD or a USB key. This distribution is usually downloaded free of charge from the official sites of the respective distributions.

While finding the software is very easy, the biggest challenge is choosing the best distribution that will best suit your computer. If you ask the question directly in a search engine "Which Linux distribution to choose? And then read the comments of the sites in response, you will quickly get confused and not get the right answer as expected.

Unfortunately, we can't help you decide on the best distribution for you because we do not know the type of computer you're using and neither can we decipher everybody's need.

However, I recommend that you test some of the most fashionable distributions to decide on which option will be best for you. This list will be highly debatable and certainly different for others.

Even though the Linux community is trying to offer you the best possible product, unfortunately, it does not always have the opportunity to obtain all the necessary information from the manufacturers of computer components to develop the drivers that accompany these components and in rare cases the selected distribution only partially work with your computer.

For example, for unrecognized WIFI cards or graphics cards that were recently developed, it is possible that no driver exists to make them work properly.

However, the big advantages of Linux is that you can download it as many times as you want and you can create as many LiveDVD or LiveUSB as you want with the distribution of your choice.

The choice of distribution depends on some requirements: For instance:

- Is your computer old or new?

- Is your computer rather powerful or not? (Processor, RAM, graphics card, hard drive, etc.)

- Do you have a preference for the office layout? (Different desktop environments are offered, Unity, KDE, Mate, LXDE, XFCE, Cinnamon, etc.)

In the case of a very old computer, one will choose a distribution which proposes 32 bits, for the recent computers one will take preferably of 64 bits.

If your PC is quite powerful, you will have little to worry about whether the desktop environment is light or not. KDE and Unity can consume more resources than XFCE or LXDE.

The best way to determine which office is right for you is to try a few. However, for beginners, I recommend the following divisions to start with

- Ubuntu with all its variations (Mate, Budgie, XFCE, KDE, etc.)

- Linux Mint and its variants (Cinnamon, Mate, XFCE, KDE, etc.)

- Zorin

- Mageia

- PCLinuxOS

How To Install

- Choose the Linux distro you want to use
- must first start by installing the ISO image
1) Once the ISO image is downloaded, it should normally be found in the "Download" directory of your hard drive.
2) Put Linux on a USB key or DVD
3) We must now place this image on a DVD or a USB key.

Depending on the choice between 32 or 64 bits, various tools are available on the internet for free for a USB key or a DVD. UnetBootIn may be suitable for 32-bit and Rufus or other distributions for 64-bit distributions.

4) Next, you will have to burn the ISO image and put the distribution you want on the USB key.
5) As soon as our USB Key or DVD is ready, you can instruct your computers to start the installation process.

For old computers, this is done by changing a parameter in the bios, for recent computers, it is done by pressing a key on the keyboard when starting the PC.

Install Linux easily

Once you have arrived in front of the office of the chosen distribution, you can find an icon that will be used to perform the installation of Linux on the hard drive.

Attention, before installing you should know that, the installation of Linux on the entire hard disk DELETE all data on the disk. So either your computer is specifically reserved for Linux installation, or you have made a full backup of your disk.

Fortunately, the installation process is the same for all distributions. The only option that differs is in the case of Linux Mint where you need to clear the disk before installing.

Chapter 4 – Linux Features, Components And Architecture

\
Linux operating system is one operating system that is free, easy, and fast to use. Its powers both servers and laptops worldwide, the Linux operating has several features, but the outstanding ones include the following

Portability:

Linux programs can be installed on any hardware; this is not necessarily about its size; this means that its software can work on different types of hardware in the same way. Linux kernel and other application programs support its installation

Freely available:

Linux source code is an open-source, and it's a community-based development project. Linux is still evolving, and multiple teams are working in collaboration to boost its capability.

Communication:

This operating system has a perfect feature for communicating with other users. Users can easily exchange program, mail, or data between two or more computer network or within a network of a single central system through these networks.

Made for multi-users:

Linux operating system is a multiuser operating system, that several users at the same can have accessibility to system resources like the RAM, Application programs, and even the memory.

It's multiprogramming:

Linux operating system can be used to run multiple applications at the same time, and it will remain active.

It is safe and secure

There is no fear of using Linux OS because it offers users security using authentication features such as encryption of data, password protection, and controlled access to specific files. The three concepts used with Linux for security are authentication, this is about claiming the person that one is and assigning such a user a password and login name for easy access, authorization Linux issues access limits to users, there's the read, write and execute permissions for each file and the OS will decide who access, modify and execute such a file. Encryption this is about encoding one's file into an unreadable format, so the OS does this to help keep user files and secrets safe.

Shell:

It offers an excellent interpreter program which can be used to carry out commands of the Linux OS. Additionally, it can be used to do other types of operations and call application programs.

Offers well-arrangedd file system

The Linux operating system provides a perfect file system whereby the system files and the user files are hierarchically organized.

Supports custom national keyboards:

Linux OS is used around the world, and as such, it is available in several languages, so it can be used with most of the national and customized keyboards.

Supports Application installation;

This OS has its own software repository that users can easily download and install thousands of applications by just issuing a command in the Shell or Linux terminal. Linux can as well be used to run windows application if desired.

Live USB and CD feature:

Linux distributions have the live USB and CD option whereby users can try or even run the Linux OS without even installing it on their system.

It has graphical user interface:

Linux OS is not only a command-line OS, but it also has packages that can be installed to make the whole OS graphics-based as windows.

Components and architecture of Linux operating system

The core components of a Linux system include the following:

The Kernel:

Kernel can be said to be the central component of the system that communicates directly with the hardware. The kernel helps to allocates system resources such as hard disk, external devices, processor and memory to the programs that are running on the computer. Kernel aid in separating each program from one another, so that when a program encounters an error, other programs aren't affected. Though users don't need to worry about the kernel in use it's on record that some versions of kernel perform better with specific hardware or software.

Daemons;

A daemon is another excellent component of the Linux OS; it is called Daemons because they are mostly invisible to the user and run things silently. They help processes like :

1. Responding to connected USB devices

2. Configuring network connections for users
3. managing files system
4. And also managing user's login.

Shell:

The shell, otherwise known as the command line aid implements a textual interface that enables one to run programs and control the system by entering commands from the keyboard. Making one system do most things would be challenging without the help of the shell or a desktop environment. There are various shells for Linux, each having different features. Some Linux systems use the bash which is the Bourne again shell, and Linux shells support multitasking which runs several programs at a go.

Devices as files

As files can be written and read in the system, so can the computer send and receive data, so because of this Linux operating system now represents the devices connected to the system as files in the directory. The fact remains that these files can't be moved or renamed and they are not stored on any disk. This is all about application programming.

The X window server

The X window server component is a graphical replacement for the command shell. It aids in drawing graphics and also processing input from the mouse, keyboard, tablets, and other devices too. This component is network transparent, which means that it allows one work in a graphical environment which will be on one personal computer or on another remote computer which one can connect across a network. The most used X server today is X.org. Some graphical programs mostly need the X server to run, which means they can be used with any desktop environment or window manager.

Window manager

The window manager component is a program that helps communicates with the X server. The real duty is managing windows. This component helps in bringing a window to the front when it is clicked, drawing the window borders, hiding it to minimize its program and moving it on the screen too. The trending window managers programs include Compiz fusion, Xfwm, KWin, and Metacity.

File system

The file system is another core component of the Linux operating system. These are the several file systems that Linux –based distributions use. These files are squashFS, BTRFS, NILFS, VFS, and EXT3/4.

The hard drive of the computer has a simple interface, and it only accepts commands like adding block block no and put it in memory address, now if one wanted to edit a piece of text and want to have it saved it on the disk, then using block numbers to identify fragments of the data like the text one needed will be awkward.

The problem will first be telling your program where to save the file using the raw block numbers, and again you have to be sure that the blocks are not used for storing other files like music collection, family photos or maybe your system's kernel.

To resolve this issue, file system was introduced. These files are organized in groups called directories. Each of the files is commonly identified through a path which explains its place in the hierarchy of the list. So in Lin,ux we have the root directory which is the top-level directory. The root directory always has a small number of subdirectories. There are:

- Bin
- Home
- Media
- TmpUsers
- Var
- and bin

Linux Architecture;

The Linux OS works correctly on four layers

which include:

- the hardware
- kernel
- shell
- and the applications

Applications/utilities

These are the utility programs that run on the shell. These consist of any application which includes text editor, web browsers, media player, etc.

Shell:

This is the interface that does the duty of taking input from users and sending instructions to the Kernel. Shell again takes the output from the kernel and sends its result back to the output shell.

Hardware:

The hardware consists of all the physical devices attached to the system, which includes the RAM, CPU, Hard disk drive, Motherboard, and lots more.

Kernel;

This is like the central component for any operating system which directly interacts with the hardware.

Chapter 5 – Comparison Between Linux And Other Operating system

Even though Linux operating system can co-exist easily with other operating systems on the same machine but there are still the difference between it and other operating systems such as windows OS/2, Windows 95/98, Windows NT and other implementations of UNIX for the personal computer. We can compare and contrast the Linux and the other operating system with the following points.

Linux is a version of UNIX:

Window NT and window OS/2 can be said to be multitasking operating system just like Linux. Looking technically at them both Windows NT and windows OS/2 are very similar in features like in networking, having the same user interface, security, etc. but there is not a version of UNIX like Linux that is a version of UNIX. So the difference here is that Linux is a version of UNIX and as such enjoys the benefits from the contributions of the UNIX community at large.

Full use of X86 PROCESSOR:

it is a known fact that Windows such as windows 95/96 cannot fully utilize the functionality of X86 processor but Linux operating system can completely run in this processor's protected mode and explore all the features therein which also includes the multiple processors.

Linux OS is free:

Other operating systems are commercial operating system though windows are a little inexpensive. Some of the cost of these other operating system is really high for most personal computer users. Some commercial operating system cost as high as a $1000 or more compared to Linux that is free. The Linux software is free because when once one can access the internet or another computer network it can be downloaded free to be installed. Another good option is that the Linux OS can be copied from a friend system that already has the software.

Runs complete UNIX system:

Unlike other operating sys,tem one can run a complete UNIX system with Linux at home without incurring the high cost of other Unix implementations for one's personal computer. Again there are tools that will enable Linux interact with Windows, so it becomes very easy to access Windows files from Linux.

Linux OS still does much than Windows NT:

Though more advanced operating systems are still on the rise in the world of personal computer like the Microsoft Windows NT that is trending now because of its server computing but can't benefit from the contributions of the UNIX community, unlike the Linux OS. Again Windows NT is a proprietary system, the interface and design are owned and controlled by one corporation which is Microsoft so it is only that corporation or Microsoft that may implement the design so there might not free version of it for a very long time.

Linux OS is more stable:

Though Linux and other operating systems such as Windows NT are battling for a fair share of the server computing market. The Windows NT only has behind it the full support of the Microsoft marketing machine but the Linux operating system has the help of a community which comprised of thousands of developers which are really aiding the advancement of Linux through the open-source model.

So looking at this comparison it shows that each operating system has its strong and weak point but Linux is more outstanding than another operating system because other operating systems can crash easily and very often especially the Windows NT while Linux machines are more stable and can run continuously for a long period.

Linux as better networking performance than others:

Linux OS can be said to be notably better when it comes to networking performance, though Linux might also be smaller than Windows NT it has a better price-performance ratio and can compete favourably with another operating system because of its effective open source development process.

Linux works better with other implementations of UNIX:

Unlike the other operating system which can't work with other implementations of UNIX, this is not same with Linux OS. UNIX features and other implementations of UNIX for personal computer are similar to that of Linux operating system. Linux is made to supports a large range of hardware and other UNIX implementations because there is more demand with Linux to support almost all kind of graphics, brand of sound, board, SCSI etc under the open-source model.

Booting and file naming:

With Linux OS there's no limitation with booting it can be booted right from logical partition or primary partition but with another operating system like the windows there is the restriction of booting it can only be booted from the primary partition and Linux operating system file names are case sensitive but with others like the Windows it is case insensitive.

Linux operating system is customizable:

unlike another operating system mostly with Windows the Linux operating system can be personalized, this is to say one or a user can modify the code to suit any need but it is not same with others. One can even change Linux OS feel and looks

Separating the directories:

With Linux, OS directories are separated by using forward slash but the separation of windows is done using a backslash. And again Linux OS uses the monolithic kernel which naturally takes more running space, unlike another operating system that uses microkernel which consumes lesser space but at the same time its efficiency is a lot lower than when Linux is in use.

Chapter 6 – Linux Tips and Tricks

.There are thousands of command lines which makes it possible to execute tasks in no time. Rather than flipping through pages one after the other, a single command can help you get what you want in few seconds.

On Linux, you musy note that some windows shortcut key may perform different function entirely. Hence, don't expect Ctrl + S to help you save items on Linux.

- ls

This command is used to list the files in a folder. You can use "afor" to list hidden files, "-lfor" detailed list, and" Rallows" to view subfolfilename

change directory, the command is used to navigate within existing file system. For example cd /var/loggo to the logs folder. This is effective regardless of where you are since we put the start slash, which indicates that it is an absolute address.

To navigate within directories, there are two very useful shortcuts to know. cd ~ leads to the directory of the current user (/home/user/most of the time or /root/if you are root) and cd -returns to the previous path.

- pwd

print working directory. This command simply displays the absolute path of the folder you're presently in.

- clear

This is used to clean your terminal window relegating all the text above and leaving you with a clean window. The keyboard shortcut ctrl+ l does the same thing.

- ctrl + s

Stop the display, this is very useful when you mistakenly input a wrong command. This command undoes the previous command allowing you to correct the mistake.

- ctrl + d

This is used to disconnects a session or terminal properly.

- **ctrl + k**

Deletes all text after the cursor and saves it to the clipboard. ctrl+ u delete command from the cursor to the beginning of the line. For insta,nce if your cursor is placed at the end of the text, this key will delete the entire line.

- **ctrl + y**

This is used to paste text copied from the clipboard

- **ctrl + r**

This command allows you to search the command history. Usually, you go back to commands already typed using the top arrow key. well with ctrl+ r, you can carry out a search in this history, do ctrl+ r, then type a piece of the command which you want to search.

- !!

In line with the practical bash shortcuts, the double exclamation point allows you to launch the last command again.

for

This is certainly the most complex command of this section, especially for beginners. for is a loop instruction. A loop is used to execute an action several times, on all the elements of a variable. For example, we can very easily rename all files in a directory to replace spaces with hyphens or comma with full stop.

ctrl + z

On windows, this is used to undo an action but on Linux, this is used to pause the current process.

bg

This is used to escape from a process that is paused in the background.

fg

This is used to resume a process in the background (if several are running at the same time, fg %n°).

at

This is used to program an event to run an hour later eg at 18:22 or at now

atq

This is used to list pending tasks.

atrm

delete task.

sleep

this command allows you to pause between the execution of two commands. Example: touch gt.txt && sleep 10 && rm gt.txt

crontab

crontab is actually a command that reads and modifies a file called "crontab". Here are the most common options:

26) e: modify the crontab,
27) l: displays the current crontab,
28) r: remove your crontab. Attention, the deletion is immediate and without confirmation!

sudo

This is used to execute a command as root.

sudo su

passes root and saves it.

chmod

change the rights on a file or folder. It modifies

users access to a file.

Chown

changes the owner of a file/folder (can only be used in root) -R option for recursion.

User Command

add user

To add a user.

password

change the password of a user, eg passwd roger.

delusion

delete a user

addgroup

create a group.

usermod

modify a user (options: -lto change the name, -gto assign a group to it, and -Gto assign several groups (separated by commas)

del group

delete a group.

groups

check-in which groups is a user added.

chgrp

change the group that owns a file (equivalent to chown user: group).

CPU and RAM usage

free

indicates the space occupied by files and the remaining free memory.

load

Displays the CPU load in the form of a graph.

ps -ef

view all launched processes. Alternatively, one can use the BSD syntax ps aux.

ps -ejH

display process in tree.

Ps -u

list the process launched by a given user eg ps -u buzut.

top

the activity of the system in real-time: load, RAM, SWAP process ... top has the advantage of being installed almost everywhere.

htop

it is an improved version of top, a little more graphics, the information is clearer and it is possible to sort/order the display according to certain criteria.

glances

Similar to top and top, glances are the dashboard of your machine as it brings together at a glance all the important metrics: CPU, load, ram, swap, i / o disks, disk filling.

iotop

in the lineage of *top, here's top which, as its name suggests, provides a real-time preview of the I / O disk.

System

w

This command is used to show those connected with your system and what they're doing.

who

This is used to show who is connected to your system.

date

Shows date time-kill

kill all

quit all occurrences of a program.

reboot

Restart the operating system.

shutdown

Program a restart or stop.

power off

This is similar yo shut down except that Poweroff comes with additional option to log off or reboot the system.

halt

allows users to "shut down" the system. However, the system can remain powered on with this command (depending on the past options and the system default settings).

Last

This command shows connection history.

lsof

list open files, list open files. this command can be very useful to see which file is in use.

hostname

Displays the hostname of the machine according to what is written in the file /etc/hostname.

name

Info about the system and the gear. The -r command option allows obtaining the version of the kernel in use.

lsb_release

lsb_release -a gives all info about the Linux distribution you're using.

lshw

Give detailed information about system hardware such as ram configuration, firmware version, motherboard configuration ... With the option -short you will get a more digestible output.

The option -c is also useful in knowing the name of the network interface that is not yet configured with the system.

lsblk

List all devices connected to your hard disk.

lspci

List all PCI devices.

lsusb

List all USB devices.

sysctl

This command is used to view and configure kernel (hot) settings.

dmidecode

read the bios info.

dmesg

displays the messages in the kernel buffer.

Run multiple commands

Suppose you want to input multiple commands, no need to wait for each command to run before inputting another, you can use the symbol ';' to separate each command so they run independently. For instance: Clear; Cd; cd~

Recover Forgotten Commands

In a situation where you forgot a couple of commands you inputted minutes ago, you can use a search term to search for the commands. To do this use the following keys: Ctrl + r. Input the command again and again to view more search results.

Exit search terms

The command Ctrl + C help you reverse search terms allowing you to return to the previous terminal.

Unfreeze Linux Terminal

If you use Ctrl +S frequently to save items on windows, you could mistakenly do so on Linux. However, rather than save items, this command actually freezes your terminal on Linux. However, you can unfreeze it using Ctrl + Q.

Move To Begin and End of Command

If you're inputting long strings of command, you can navigate to the first line of Command using Ctrl + A and Ctrl + E to move to the end.

How To Reuse the Previous Command

You can use !! to recall previous commands in a new line.

Copy and Paste Commands

On Linux, you can copy command on your terminal and paste where you need it. To do this, simply highlight the command you want to copy and press Ctrl + Shift + C to copy and Ctrl + Shift+ V to paste.

Empty File Content With deleting the file

You can delete the contents in a folder while leaving the folder intact using the command: > followed by the file name. For instance : > Linux tutorial

Recall Commands

If you have long lines of Commands, you can recall Commands in a specific line using '!' followed by the line number. For instance: !23 to view command inline 23.

Shutdown Computer at A Given time

In addition to shutdown and reboot, you can also program your computer to shut down at a specific time. To shutdown Computer at 22:00, use the following command: $ sudo shutdown 21:00

mkdir

With this, you can create a folder. The operation is the same as that of the command touch. Eg mkdir Linux tutorial

cp

With this, you can make a copy of a file. The option -R allows you to make copies of entire folder at once.

mv

This option allows you to move files/folders. The mv command is used in exactly the same way as the command cp. In addition, this command also allows you to rename files and folders.

rm

remove, delete files. E.g rm Linux tutorial. The option -f forces the deletion, the option -i requests confirmation before deletion, finally the option -r allows the deletion of the files.

rmdir

remove directory, and allows you to delete a folder only if it is empty.

ln

This option helps you create a link between two files. The option -s allows you to create a symbolic link.

wc

word count, count the number of lines, words, and characters in a text file. The options are -lfor line (number of lines), -wfor word (number of words) and -mfor the number of letters. There is also the option -cto have the file size in bits. To use it, we simply provide in parameter the address of the text file:

wc also makes it easy to know how many files/folders you have in a given directory, just to redirect the output of a diverse WC: ls | and voila!

sort

This option helps you sort a text file in alphabetical order. The option -r allows to perform an inverse sorting, ie anti-alphabetical or decreasing for numbers, and the option -R allows random sorting.

Uniq

the command unit allows you to duplicate a file. Just supply in parameter the address of the file to be duplicated and the name of the new file to be created.

Backup Files

is a utility that allows you to synchronize folders. Very practical so for the backup. It's the options -areI use. -retains rights etc, -allows recursion and -for verbose mode. A small example of saving my vacation photos:

file

This command helps you determines the type of file in use regardless of its extension. All you need to do is provide the parameter of the file to be evaluated.

split

With this command, you can cut a file into smaller files (-l specify the number of lines, -b specify size in bytes [follow the size of K, M, G, T to define a different unit]).

Locate

this command allows locating a file on the hard disk. Eg locates linuxtutorial.txt. The command locate is very fast because it finds the file by consulting a database. It does not scan the hard drive directly for the file in question. The disadvantage of this process is that if the file is recent, it may not be indexed yet, and location will not be of any help.

Find

the command find is much more powerful than locate, but it is also much slower because it traverses the disc when researched. Unlike locate, find allows you to search according to the size or a date of last access?

nohup

This command allows users to start a program and holds it even after the console is closed.

cat

This command allows you to read the contents of a text file cat Linux tutorial.txt

less

similar operation cat but displays the file page by page. It is, therefore, more convenient for long files.

head

displays the header of a file, the option -n allows you to specify the number of lines to display.

tail

similar to head but refers to the "tail" of the file, in other words, this command only displays the end.

touch

The touch primary goal is to change the time frame of a file. If you make touch a file that already exists, it will update its last access and modification dates.

Make a Typescript of all elements on your terminal

With the"script" command, you can make a Typescript of everything on your Console.

Privacy of directories

To prevent other users from accessing a folder, use the command chmod

Password File

If you're afraid that a user may access your file, you can password it with the following command: Vim followed by filename.

Know When to Exit your terminal

On Linux, you can set a reminder to know when to exit your terminal. With the command "leave + time of the day", you know when to stop working.

Formatting Text

You can format text on Linux using the following string of commands: fmt + filename

Complete Last command

Using the command ! help you save time as it automatically helps you complete the last command without having to input the remaining commands manually.

Type

This command Indicates whether an element is built-in, a program or an alias.

Copy file into different directories

The command: echo allows you to copy a file into as many directories as you want with just on linc of command

Delete large files

rm command is widely used to remove files however this can become useless if the file in question is large. Instead of rm, use (>(followed by the filename) to remove large files.

Conclusion

Linux operating system is so outstanding in many ways. However, every operating system has its strengths and weaknesses Linux is easily accessible since it can be freely downloaded and installed at no cost.

Linux operating system can be said to be more reliable than another operating system since it always stable and hardly crash for a long time.

However, Linux remains a second option to most users especially beginners.

While you may find it hard adapting to this operating system if you're an avid window user, on the long run, you will discover how awesome this system is and also enhances your abilities.

To master all the trips listed here, we advise you put them into use daily. Rather than making several strokes, we advise you use these command lines to perform task better.

Finally, we thank you for reading to the end but we will be sad to see you stop here. We employ you to search other channels for tips on how to improve your experience as a user. The more you search, the more information you get and the better you become at handling this operating system.

I hope, that you really enjoyed reading my book.

Thanks for buying the book anyway!